This book belongs to

Thank you

Practice make perfect !

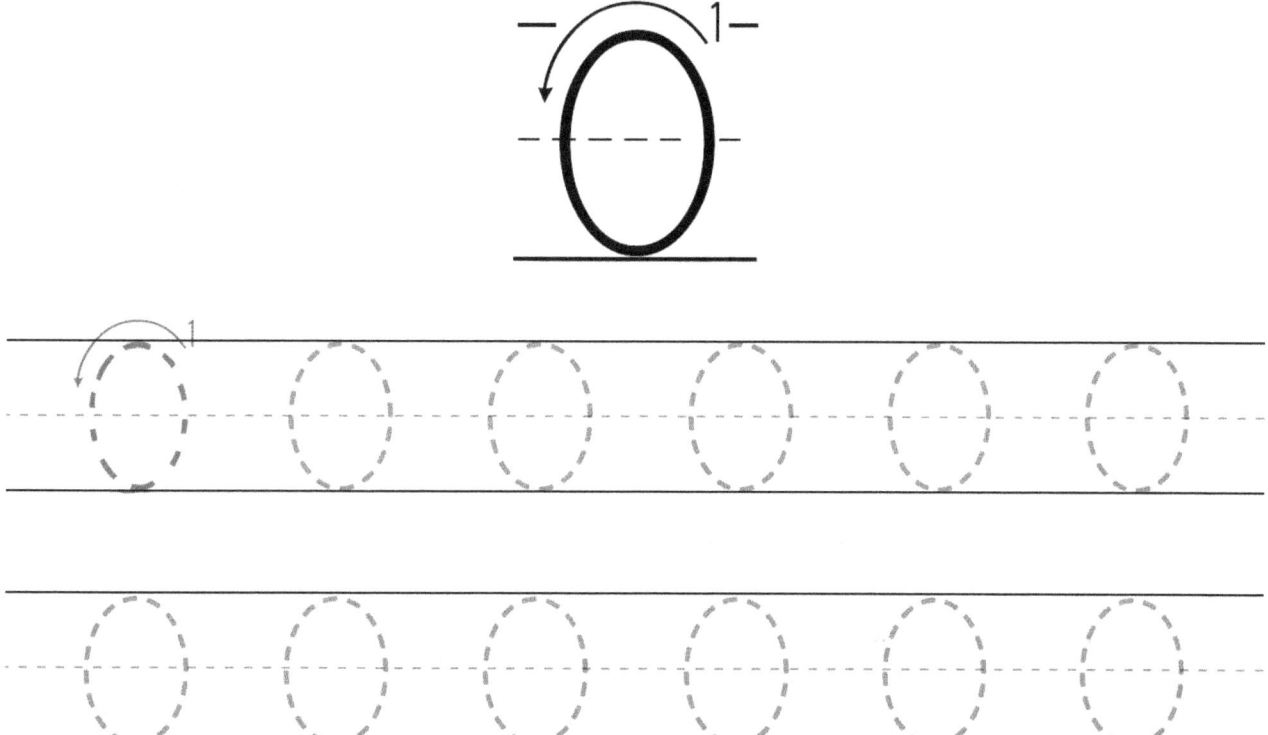

How many eggs in the basket ?

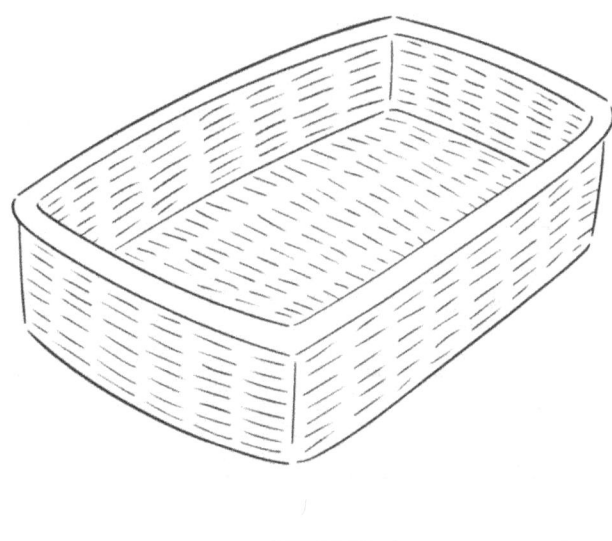

There are _____ egg in the busket.

one

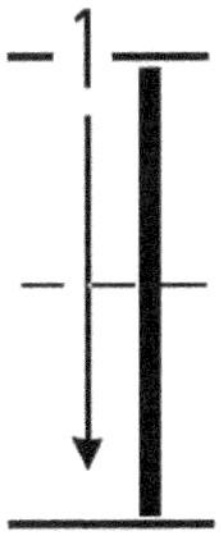

How many chickens in the picture ?

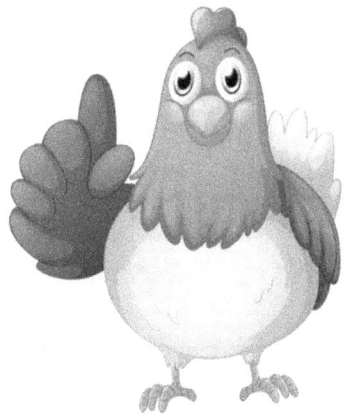

There are ____ chicken.

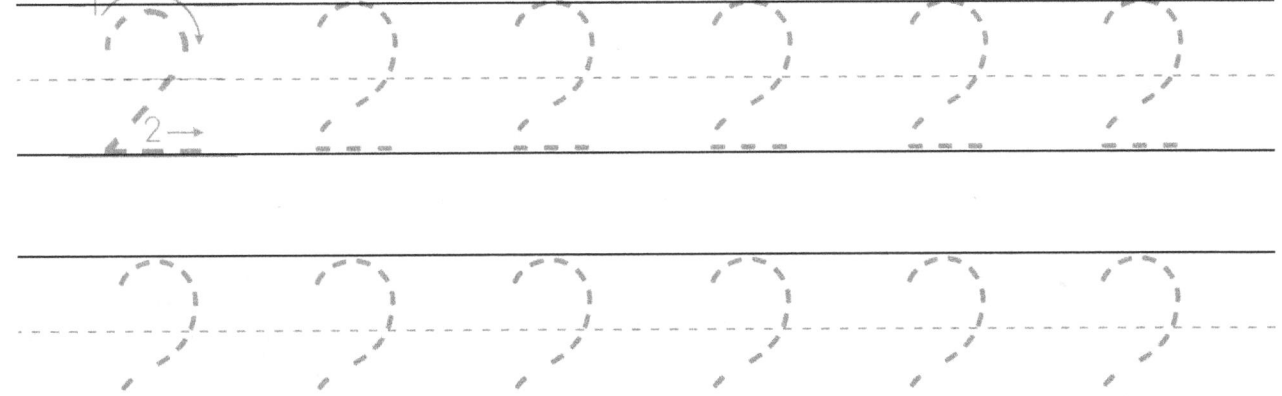

How many ducks in the picture ?

There are _____ ducks.

two

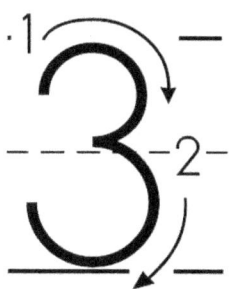

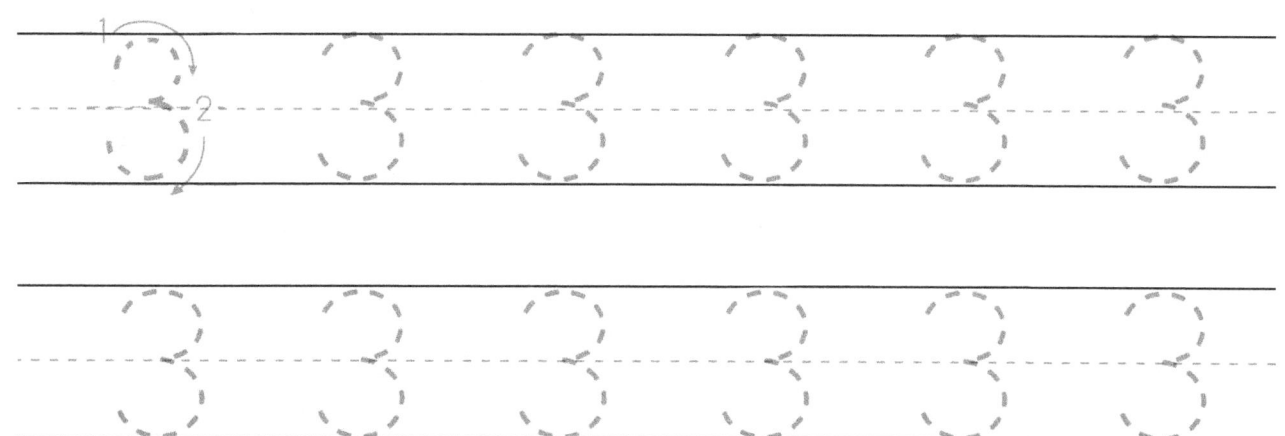

How many cats in the picture ?

There are _____ cats.

three

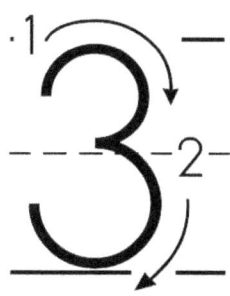

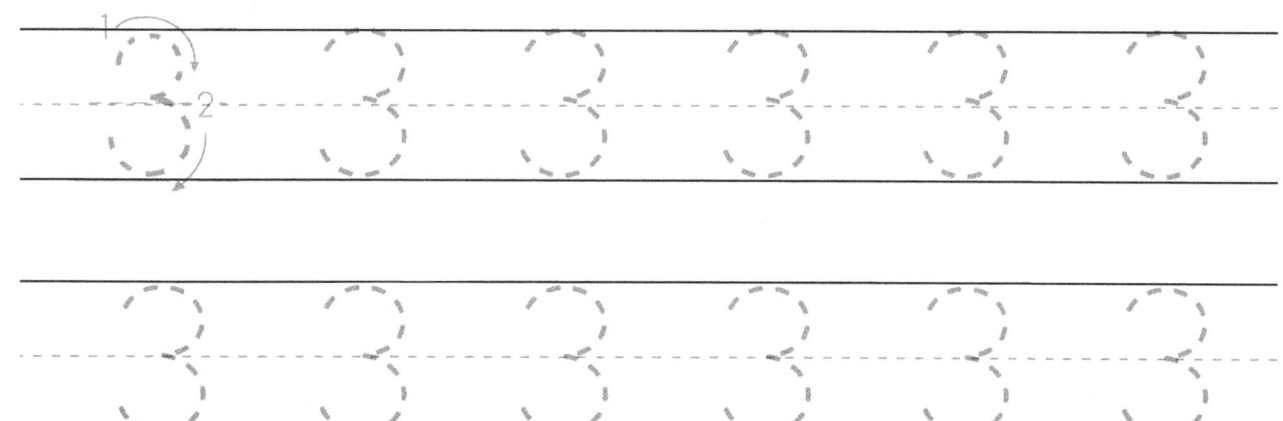

How many cats in the picture ?

There are _____ cats.

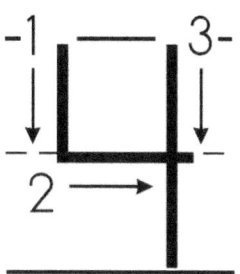

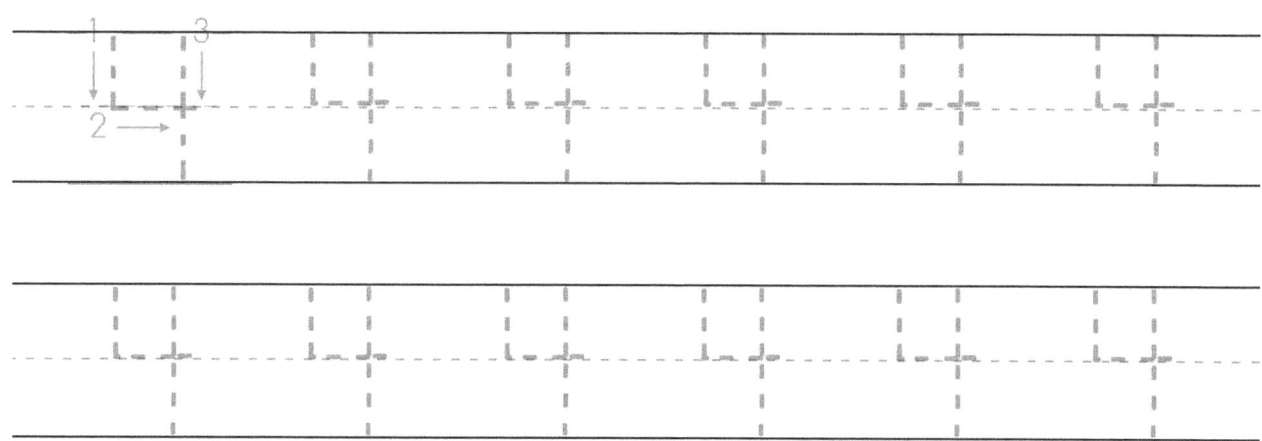

How many giraffes in the picture ?

There are _____ giraffes.

four

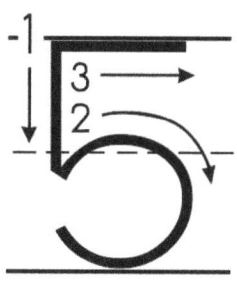

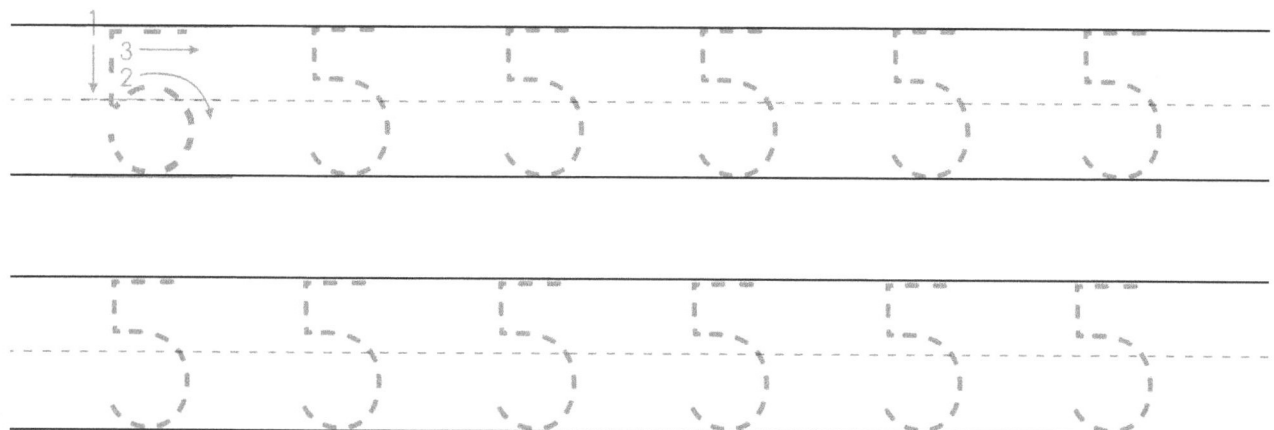

How many tigers in the picture ?

There are _____ tigers.

five

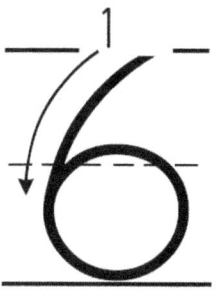

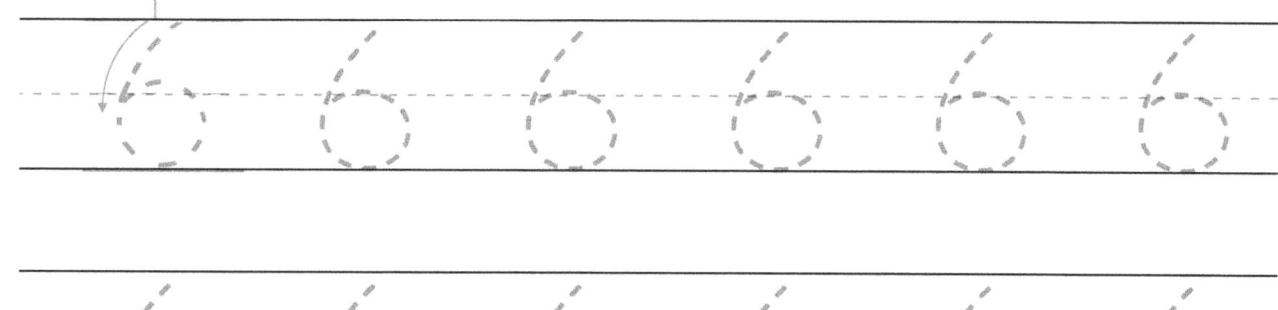

How many monkeys in the picture ?

There are _____ monkeys.

1

6 6 6 6 6 6

6 6 6 6 6 6

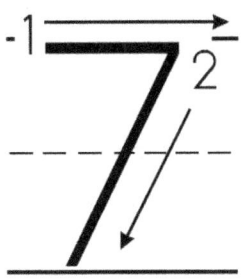

How many snails in the picture ?

There are _____ snails.

seven

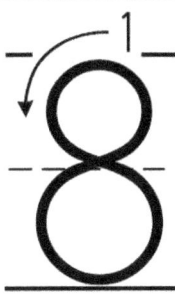

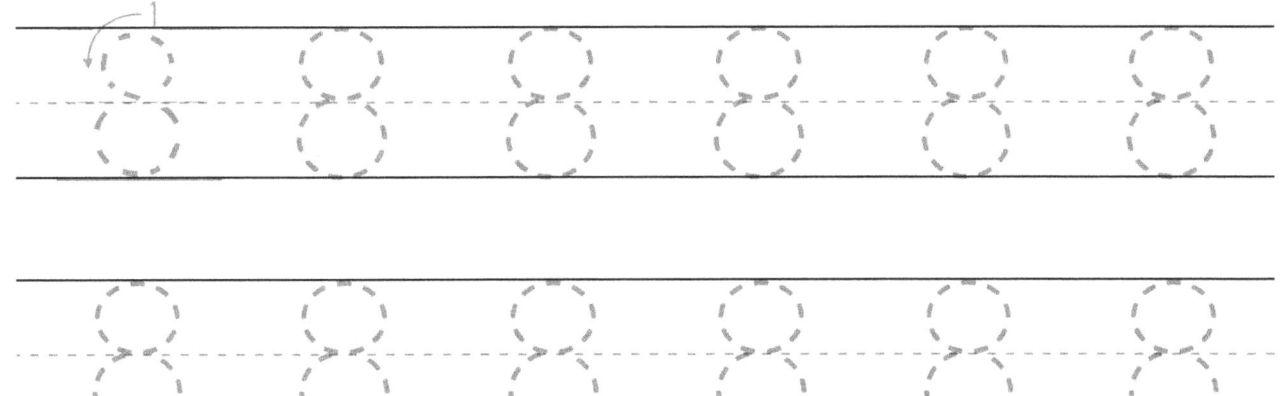

How many birds in the picture ?

There are _____ birds.

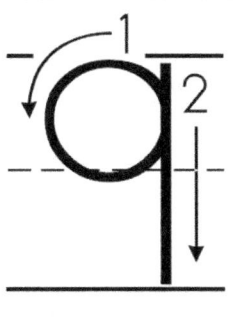

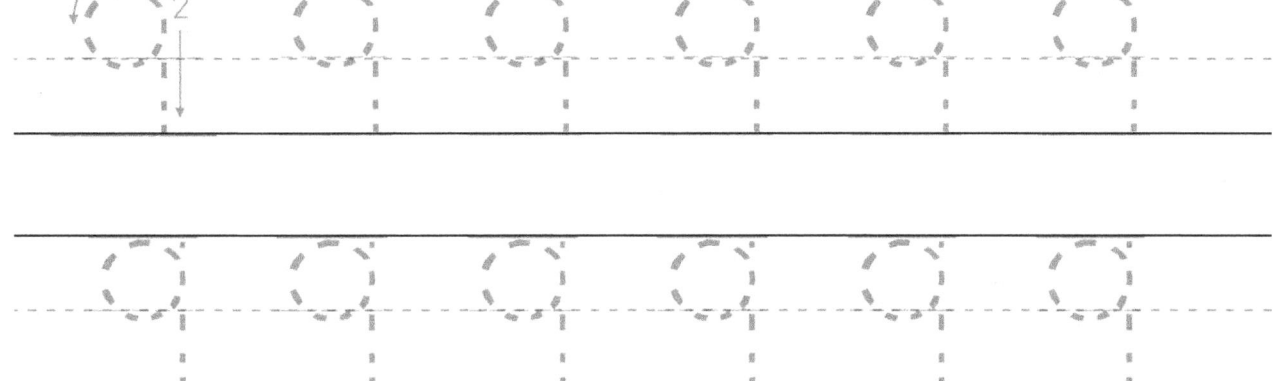

How many turtles in the picture ?

There are _____ turtles.

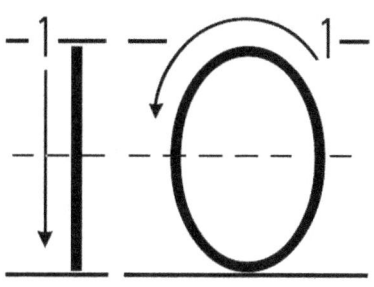

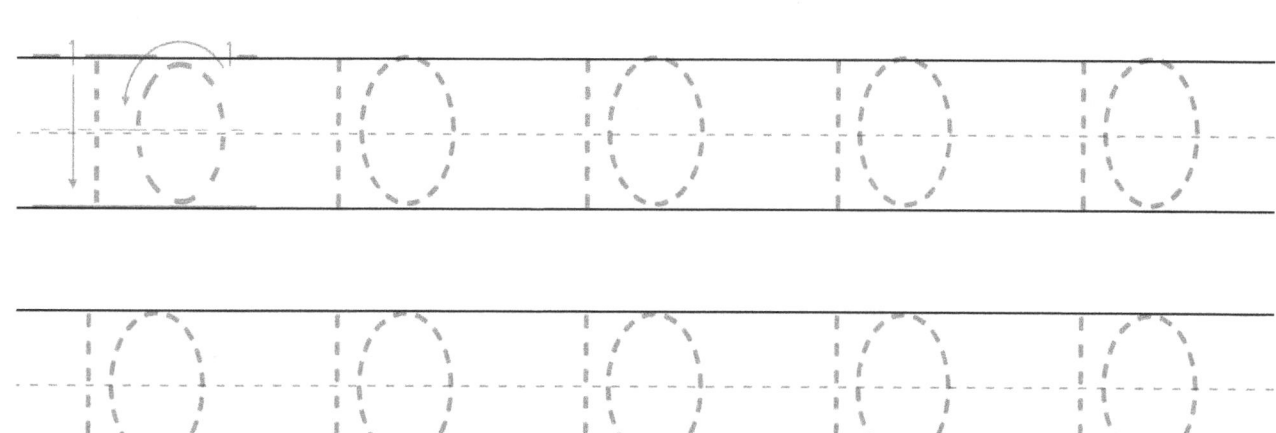

How many bees in the picture ?

There are _____ bees.

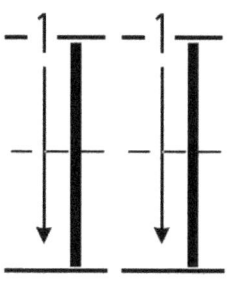

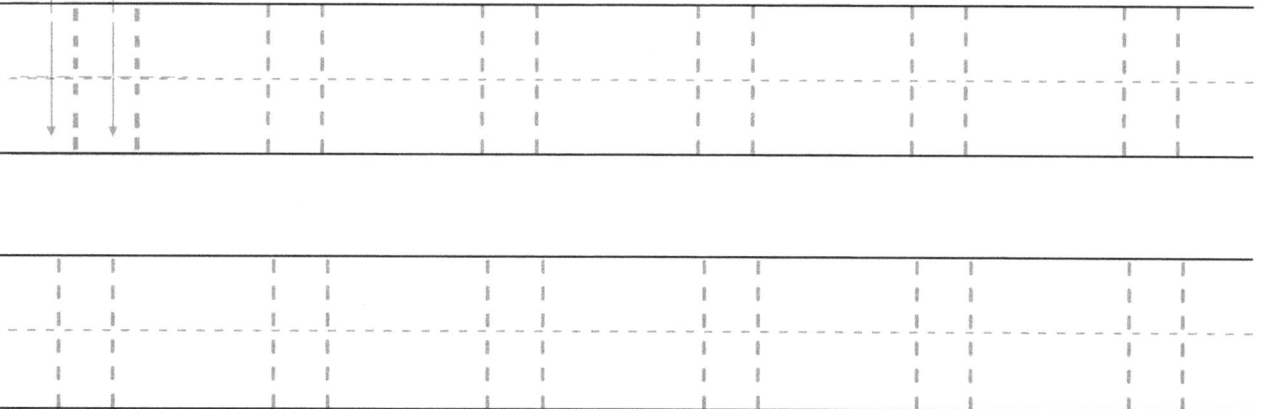

How many kangaroos in the picture ?

There are _____ kangaroos.

eleven

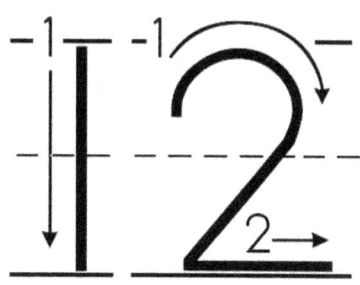

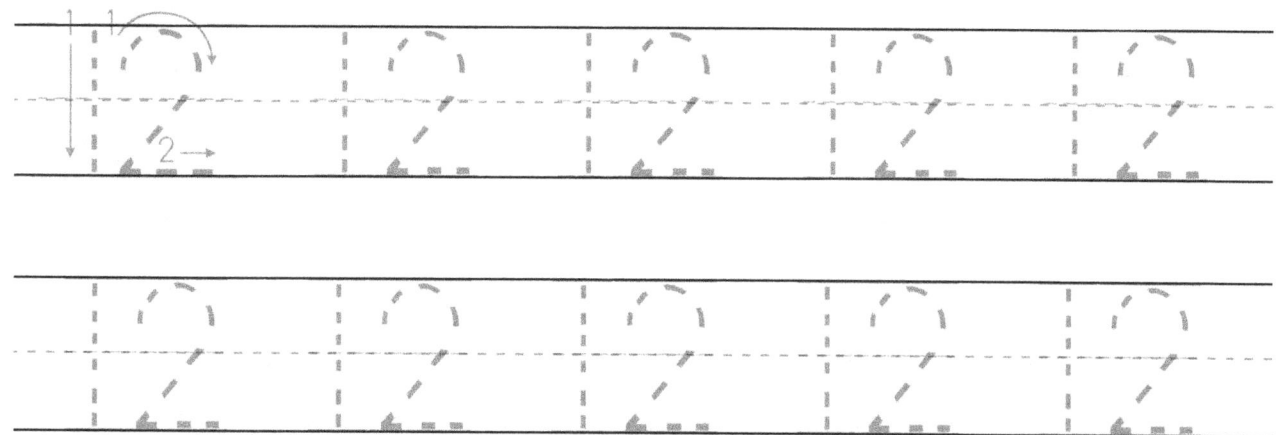

How many rabbits in the picture ?

There are _____ rabbits.

12 12 12 12 12

12 12 12 12 12

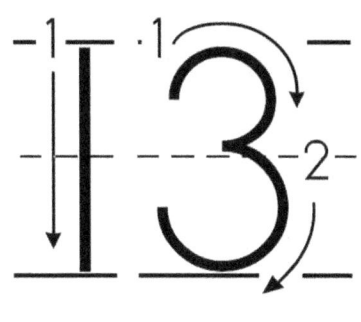

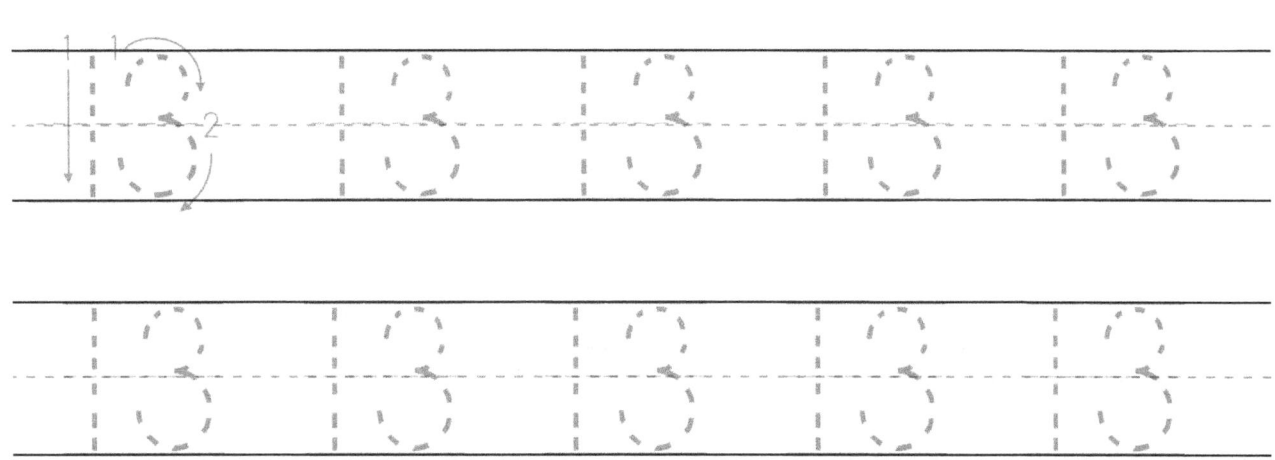

How many horses in the picture ?

There are _____ horses.

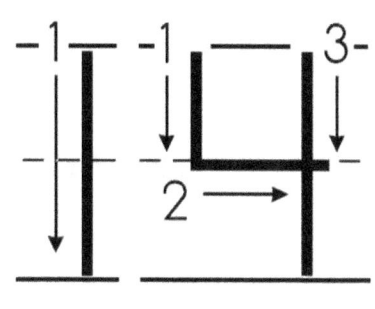

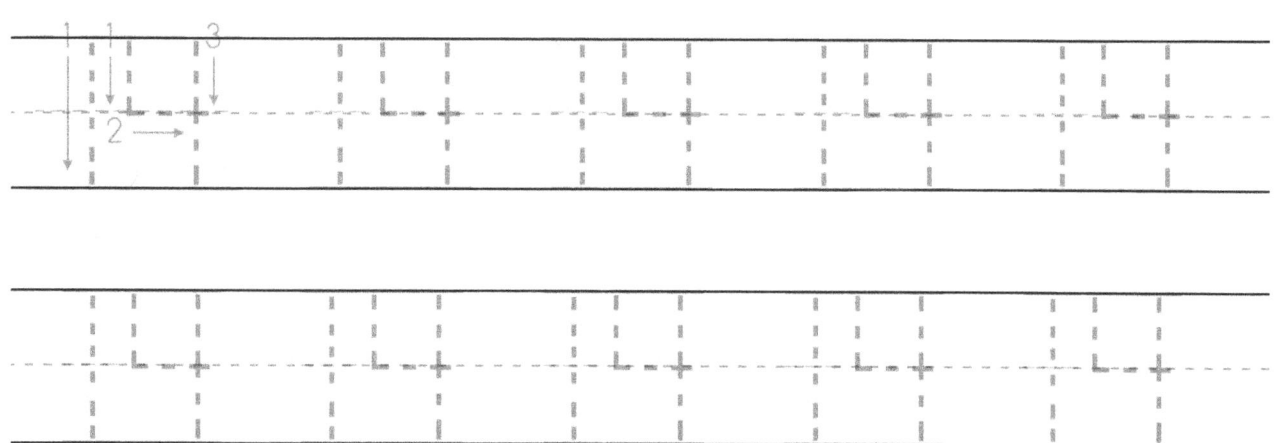

How many lions in the picture ?

There are _____ lions.

fourteen

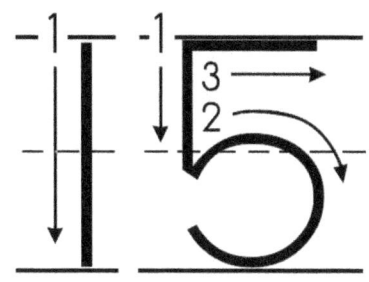

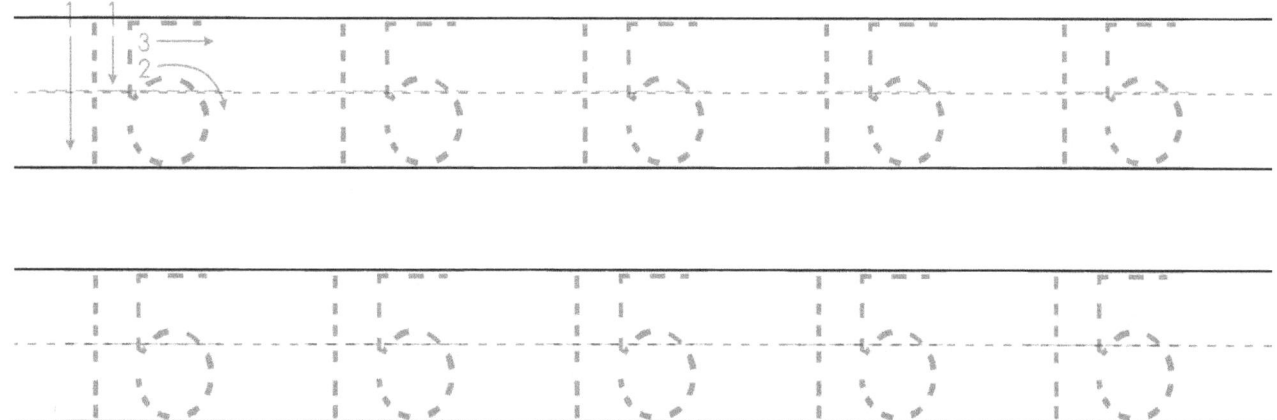

How many butterflies in the picture ?

There are _____ butterflies.

15 15 15 15 15

15 15 15 15 15

How many elephants in the picture ?

There are _____ elephants.

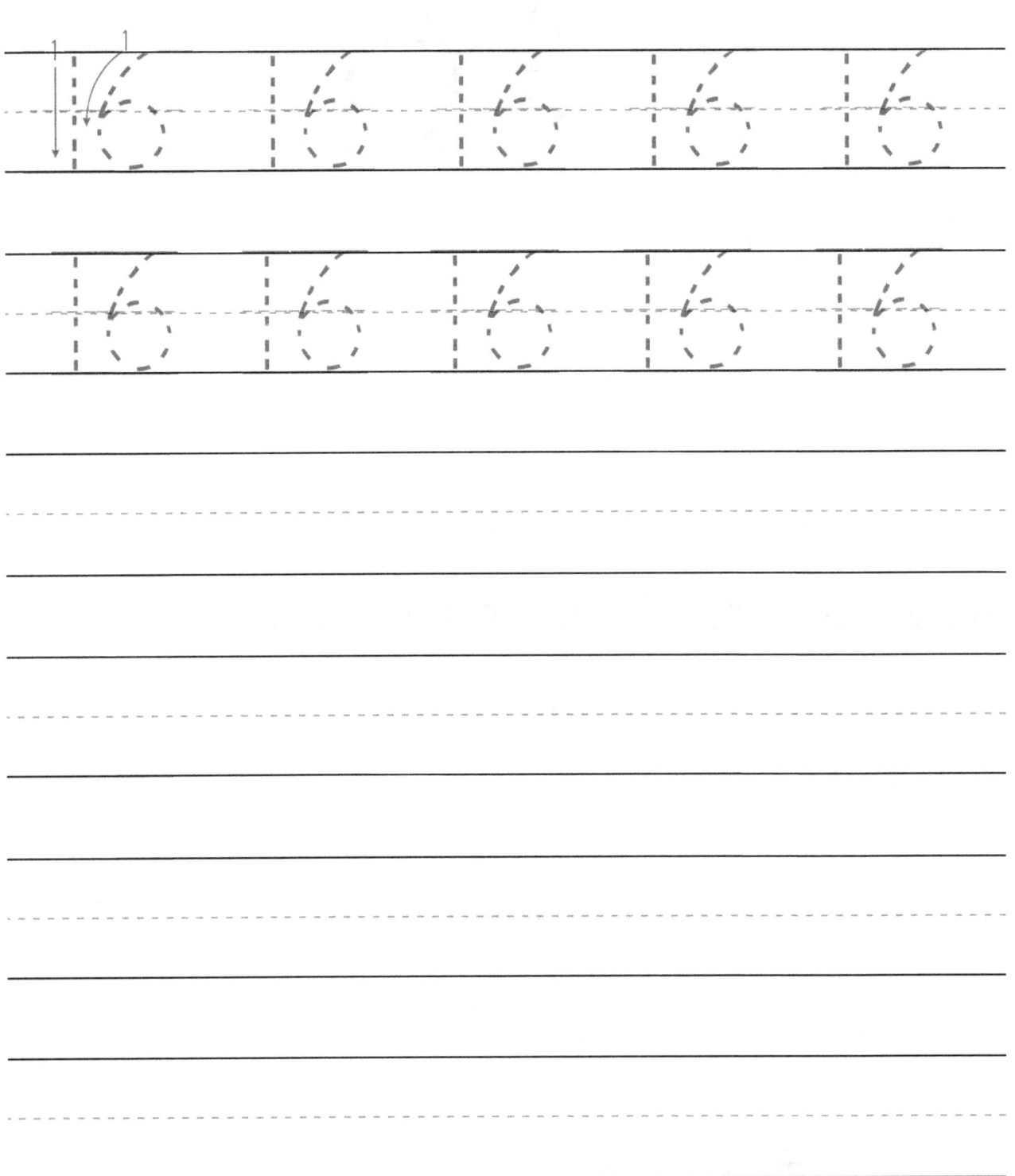

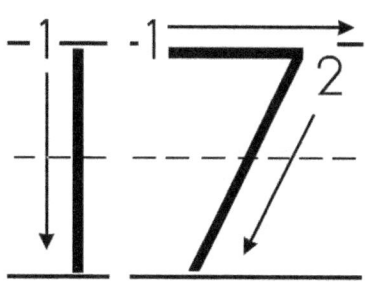

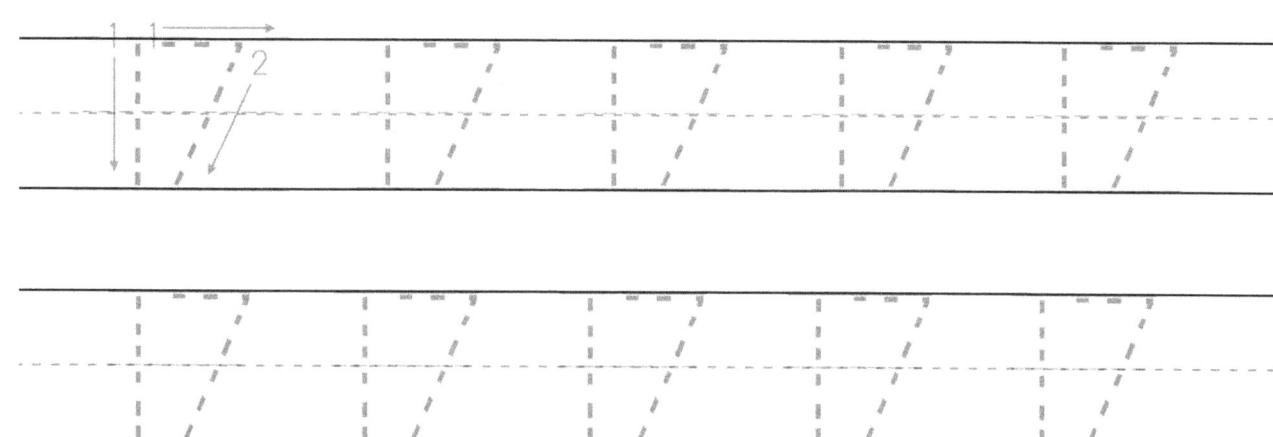

How many ostriches in the picture ?

There are _____ ostriches.

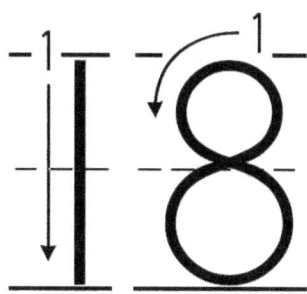

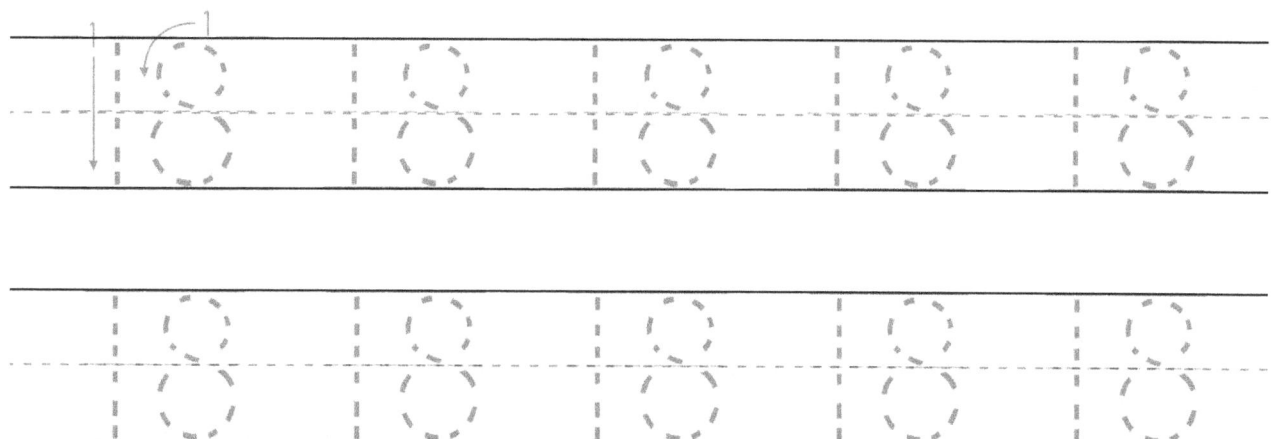

How many dogs in the picture ?

There are _____ dogs.

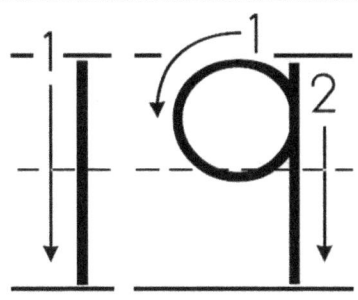

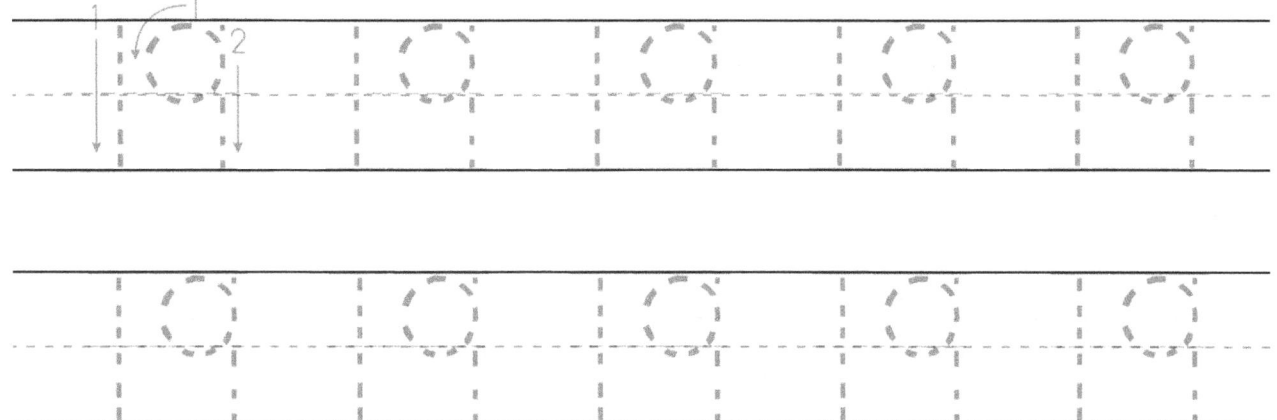

How many cows in the picture ?

There are _____ cows.

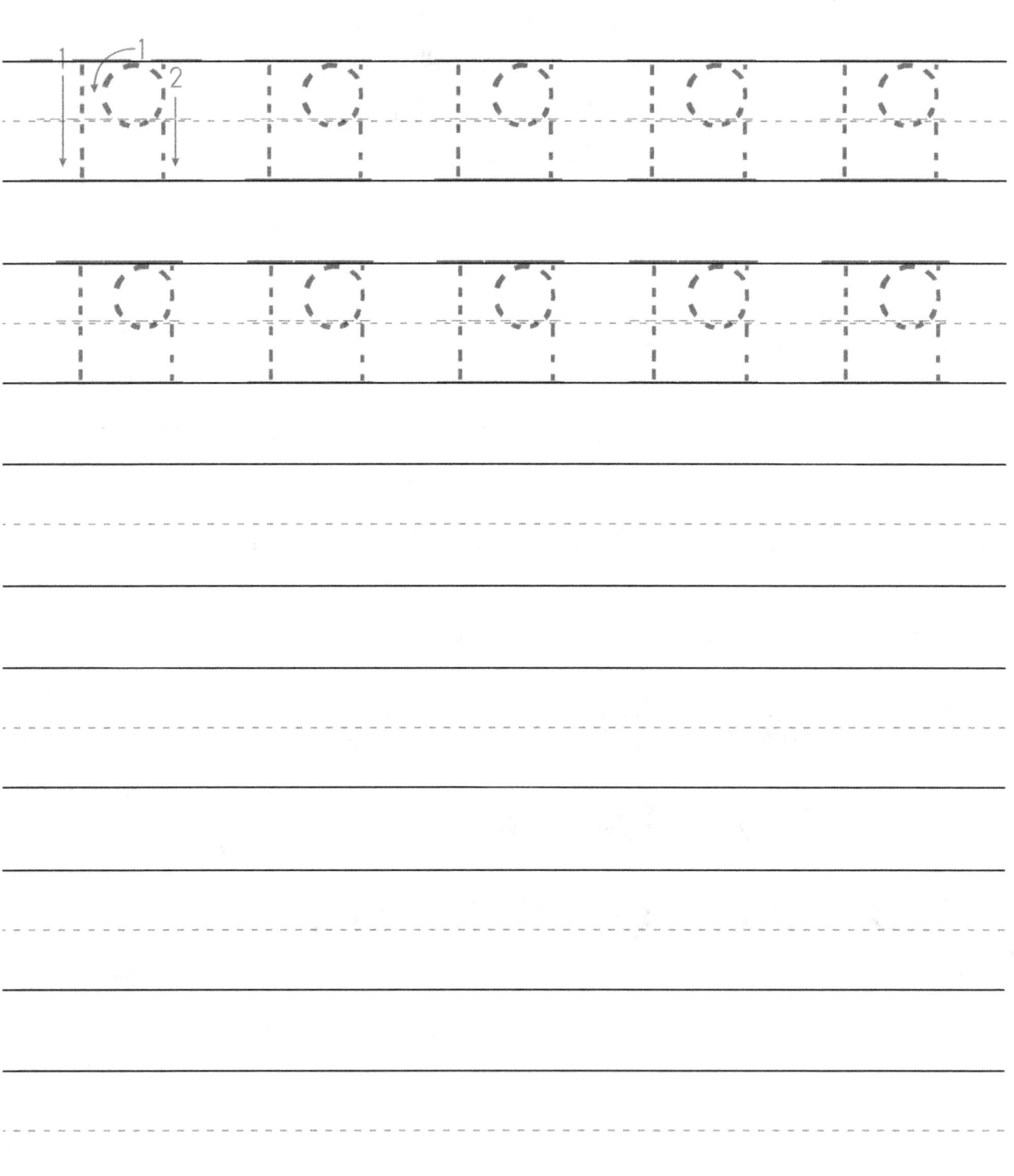

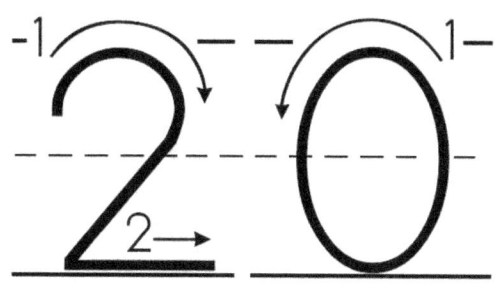

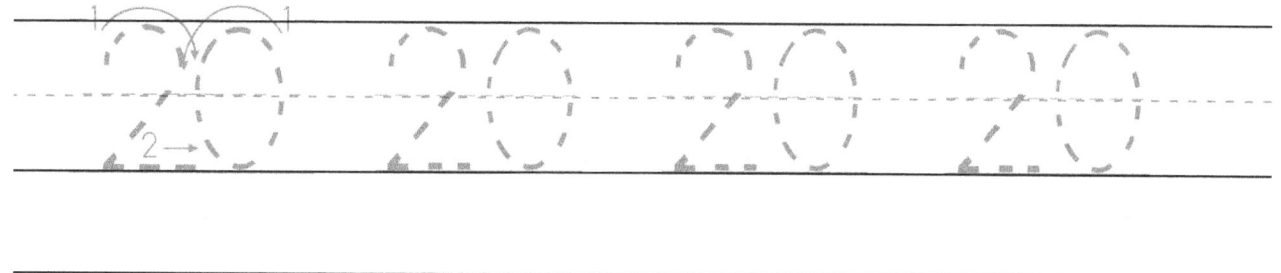

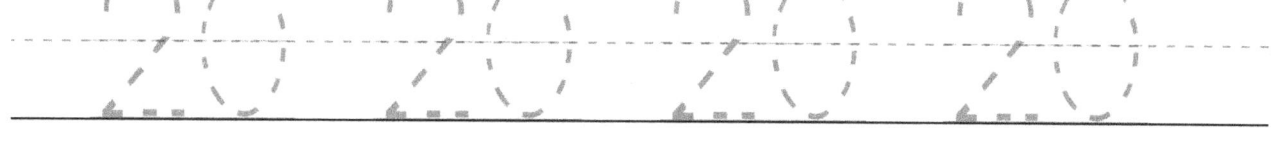

How many dinosaurs in the picture ?

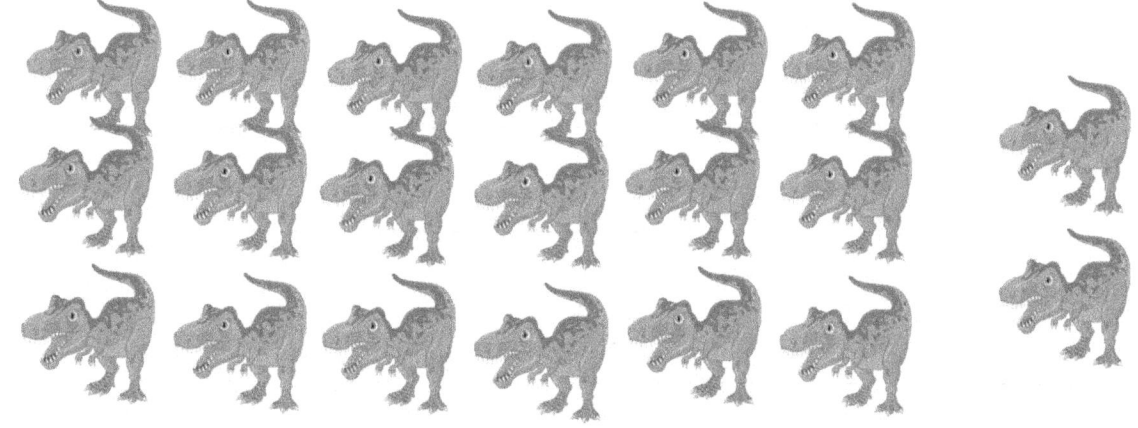

There are _____ dinosaurs.

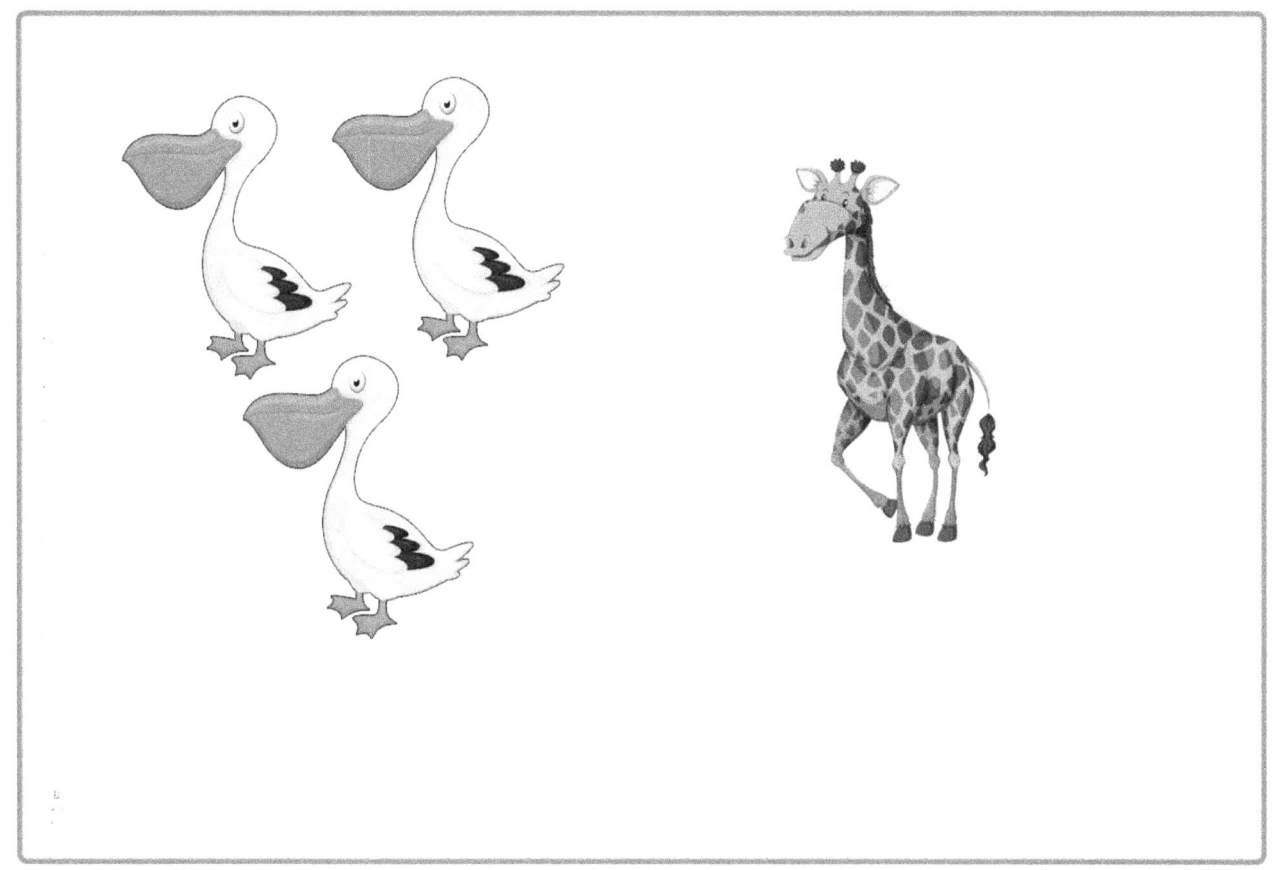

How many ducks in the picture ?

There are _____ ducks.

How many giraffes in the picture ?

There are _____ giraffe.

How many chickens in the picture ?

There are _____ chickens.

How many cats in the picture ?

There are _____ cats.

How many monkeys in the picture ?

There are _____ monkeys.

How many birds in the picture ?

There are _____ birds.

How many turtles in the picture ?

There are _____ turtles.

How many bees in the picture ?

There are _____ bees.

How many lions in the picture ?

There are _____ lions.

How many kangaroos in the picture ?

There are _____ kangaroos.

How many eggs in the picture ?

There are _____ egg.

How many dinosaurs in the picture ?

There are _____ dinosours.